DOING WORK WITH SIMPLE MACHINES

WORKING WITH WHEELS AND AXLES

RONALD MACHUT

PowerKiDS press.
New York

Published in 2020 by The Rosen Publishing Group, Inc.
29 East 21st Street, New York, NY 10010

Editor: Elizabeth Krajnik
Book Design: Reann Nye

Photo Credits: Cover, p. 10 Narudom Chaisuwon/Shutterstock.com; p. 6 feelphoto2521/Shutterstock.com; p. 7 Rolf E. Staerk/Shutterstock.com; p. 9 Polo Gtz/Shutterstock.com; p. 11 Mattia Mazzucchelli/Shutterstock.com; p. 13 Westend61/Getty Images; p. 14 Cezary Wojtkowski/Shutterstock.com; p. 15 MARGRIT HIRSCH/Shutterstock.com; p. 17 biletskiy/Shutterstock.com; p. 19 dimcars/Shutterstock.com; p. 21 iMoved Studio/Shutterstock.com; p. 22 Naypong Studio/Shutterstock.com.

Cataloging-in-Publication Data

Names: Machut, Ronald.
Title: Working with wheels and axles / Ronald Machut.
Description: New York : PowerKids Press, 2020. | Series: Doing work with simple machines | Includes glossary and index.
Identifiers: ISBN 9781538345368 (pbk.) | ISBN 9781538343654 (library bound) | ISBN 9781538345375 (6pack)
Subjects: LCSH: Wheels–Juvenile literature. | Axles–Juvenile literature.
Classification: LCC TJ181.5 M33 2020 | DDC 621.8-dc23

Manufactured in the United States of America

CPSIA Compliance Information: Batch #CSPK19. For Further Information contact Rosen Publishing, New York, New York at 1-800-237-9932

CONTENTS

WHAT'S A WHEEL AND AXLE?

Wheels and axles are found in many places, but you might not even know they're there. Ferris wheels, doorknobs, and steering wheels are all examples of wheels and axles.

The wheel and axle is one of six simple machines. A simple machine is a **device** with few or no moving parts. These devices are used to modify, or change, motion and force to do work. Force can be applied to either the wheel or the axle. When you turn a doorknob, you're applying force to the wheel. A ceiling fan is an example of force being applied to the axle.

MECHANICAL MARVELS

Some devices are made up of more than one simple machine. They're called **compound** machines. Cars, dishwashers, and lawn mowers are all compound machines.

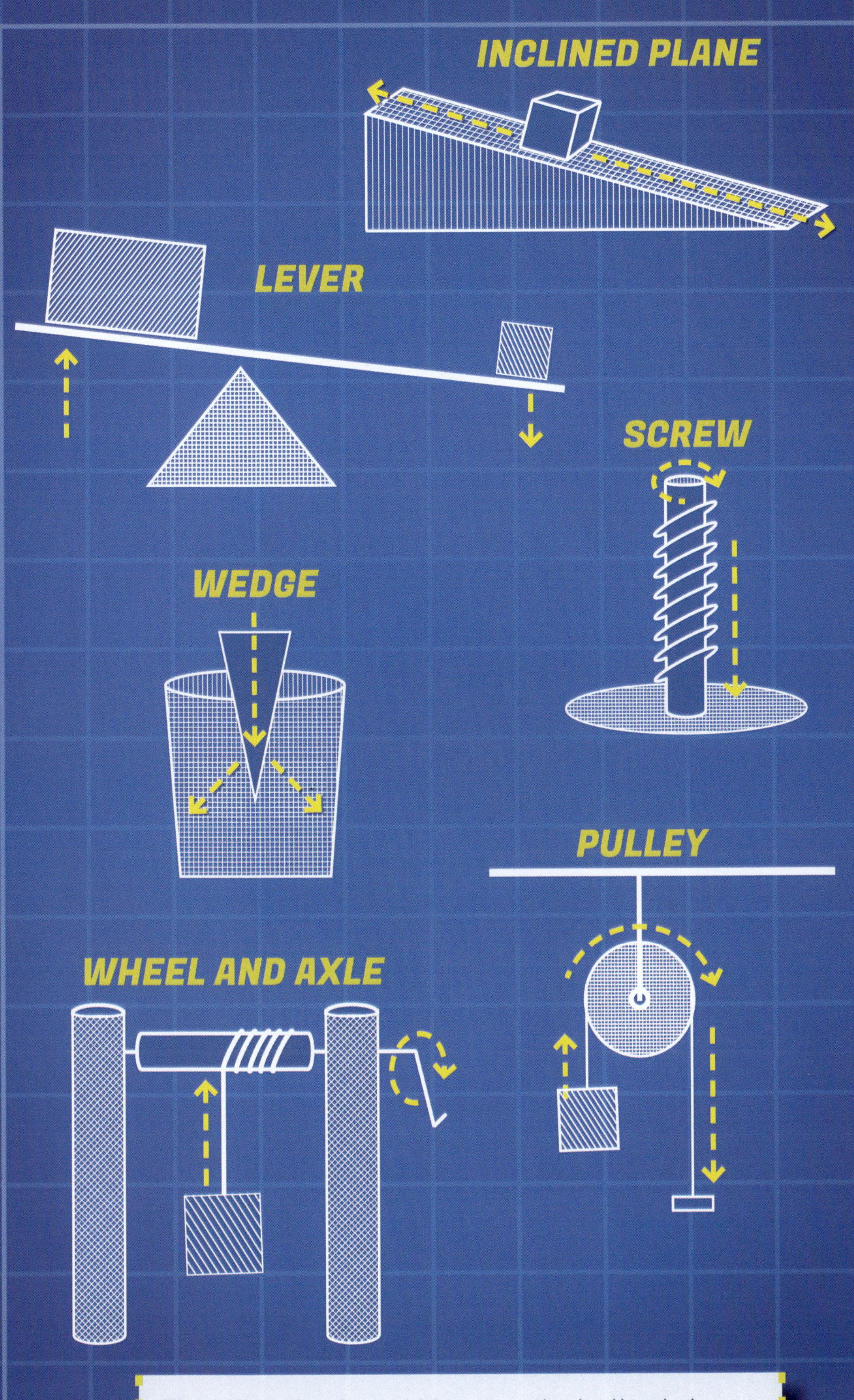

The other simple machines are the inclined plane, the wedge, the pulley, the screw, and the lever.

WHEEL AND AXLE PARTS

A wheel and axle is made up of two circular objects of different sizes that are joined at the center. The larger circular object is called the wheel and the smaller rod is called the axle.

MECHANICAL MARVELS

Some wheels and axles, such as a screwdriver, are powered by human effort. Others, such as cars, school buses, and trucks, are powered by engines.

Without its axle, a Ferris wheel wouldn't be able to spin. It also wouldn't be considered a simple machine.

To be considered a simple machine, the wheel and axle must be joined. A wheel that turns around an axle but isn't connected to the axle isn't a simple machine. Some wheels and axles may have a wheel on just one end of the axle. Other wheels and axles may be made up of wheels on both ends of the axle.

A SPECIAL SIMPLE MACHINE

Wheels and axles are special simple machines because they act like another simple machine. A wheel and axle is actually a kind of first-class lever. In most cases, a lever is a long **plank** that pivots, or turns, on or around a central point called a fulcrum.

When you apply force to the wheel, the wheel multiplies the force to make the force produced by the axle greater. When you apply force to the axle, the force makes the wheel turn a greater distance than it would with the force applied to the wheel alone. The axle multiplies the force to the wheel.

In the case of a wheel and axle, the wheel acts as one end of the lever and the axle acts as the other. The point at the center of both the wheel and the axle is the fulcrum.

DOING WORK

Wheels and axles do important work. But each part of a wheel and axle does its work in a different way. A doorknob is a wheel attached to an axle inside the door. When you apply effort to the knob, the axle turns the parts that allow the door to open.

Some cars only deliver power to either the front wheels or the back wheels. Other cars have four-wheel drive.

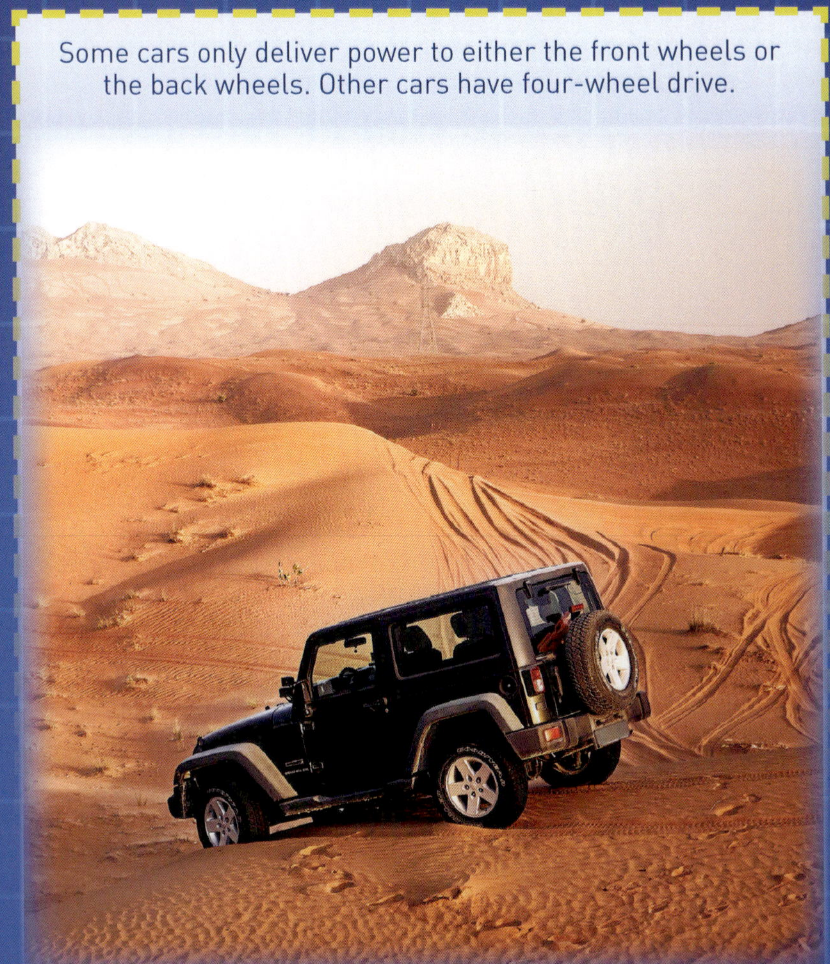

In the case of cars and other **vehicles**, force from the engine is applied to the axle. When the axle spins, it causes the car's wheels to spin. The engine's power is greater than the **friction** of the road, which allows the car to move forward with the help of the wheels and axles.

THE EARLIEST WHEELS AND AXLES

Unlike some other simple machines, wheels don't exist in nature. The people of **Mesopotamia** invented the earliest wheels around 3500 BC. They used these wheels, which were made of solid stone or circles of wood cut from trees, as potter's wheels.

A potter's wheel is a flat, round stone or piece of wood attached to a **pedal**. When the potter pushes the pedal down, an axle turns, which spins the wheel. To make clay pots, a potter needed to shape clay by hand or use a potter's wheel. Clay pots were very important in everyday life in ancient times. They were used to store food, water, and other things.

MECHANICAL MARVELS

Historians have said that the wheel is one of very few inventions that wasn't **inspired** by something in nature.

Sometime between the sixth and fourth centuries BC, people in Greece invented the wheelbarrow. This invention spread throughout the world from China to Europe.

WHEELS AND AXLES THROUGHOUT HISTORY

People have used wheels and axles to help them do work for thousands of years. Early windmills used several wheels and axles to work. The large sails of a windmill turned on an axle, powered by the wind. The axle was joined to wheels and gears within the main building.

People have put wheels on toys for thousands of years. In the 1940s, **archaeologists** found wheeled toys in Vera Cruz, Mexico, from before Christopher Columbus arrived in the Americas.

The wheels and gears turned as the sails turned, carrying the power of the wind to the millstone, which broke down corn or other grains into flour. Wheels and axles were also used to lift water from wells. These wheel and axle machines were called watermills.

USING WHEELS AND AXLES TO TRAVEL

Wheels and axles weren't used for travel until around 300 years after the invention of the potter's wheel. **Chariots** may have been the first vehicles of this sort. People had to train animals to pull wheeled wagons and carts. Then they needed to build roads that were smooth enough and long enough to be useful.

As time went on, people realized that wheels could move more quickly if they weren't solid and heavy. So, wheel makers cut long rods called spokes and used them to join the center of the wheel to its outer rim. They made the wheels out of different materials, such as wood, so they would be lighter.

Today, you can see spokes on your bike wheels. You may also see them on the wheels of the cars on the road.

SUPERFAST COMPOUND MACHINES

Compound machines are all around us. Many of these compound machines, such as the jet engine of an airplane and the propeller of a helicopter, use wheels and axles.

These compound machines are getting faster every year. Today, some cars, such as the Koenigsegg Agera RS, can go more than 275 miles (442.6 km) per hour. The Koenigsegg Agera RS is officially the fastest car in the world. This car is light and low to the ground and has a powerful engine. However, without wheels and axles, the Koenigsegg Agera RS wouldn't go very fast at all.

Koenigsegg no longer produces the Agera. Its replacement will be lighter, but it will have the same type of roof and engine.

MECHANICAL ADVANTAGE

All simple machines provide mechanical advantage, which is the machine's ability to multiply the mechanical force applied to the load. Mechanical advantage makes work easier for us. We apply force to the machine, and the machine multiplies the applied force or changes its direction. The applied force must be greater than the **resistance** of the object we're trying to cut, move, or hold together.

The ideal mechanical advantage of a wheel and axle is the **radius** of the wheel divided by the radius of the axle. However, energy is lost to friction and other forces.

A wheel and axle can increase or decrease the force applied to it. If the force is applied to the axle, the mechanical advantage is less than one. If the force is applied to the wheel, the mechanical advantage is greater than one.

EVERYDAY WHEELS AND AXLES

Wheels and axles are everyday simple machines. The handles of a faucet work like a wheel and axle. When you turn the handle of the faucet, it turns the axle to which it's attached. The axle then opens and closes the pipes that allow water to flow out of the faucet.

A screwdriver works like a wheel and axle. When your hand grips the handle of a screwdriver and turns it, the axle turns and twists the screw into or out of place more easily than you could using just your fingers.

GLOSSARY

archaeologist: Someone who studies the tools and other objects left behind by ancient people.

chariot: A vehicle with two wheels that was pulled by horses and sometimes used in battle in ancient times.

compound: Made up of two or more parts.

device: A tool used for a certain purpose.

friction: The force that resists motion between bodies in contact.

inspire: To give someone an idea about what to do or create.

Mesopotamia: The area of southwestern Asia between the Tigris and the Euphrates rivers that reaches from the mountains of eastern Asia Minor to the Persian Gulf.

pedal: A lever worked by a foot or feet.

plank: Something that supports, such as a board.

radius: The length of a straight line that goes from the center of a circle to the outside edge.

resistance: An opposing or slowing force.

vehicle: A machine used to carry goods from one place to another.

INDEX

WEBSITES

Due to the changing nature of Internet links, PowerKids Press has developed an online list of websites related to the subject of this book. This site is updated regularly. Please use this link to access the list: www.powerkidslinks.com/dwsm/wheels